AfroBiotech Conference 2019

Atlanta, Georgia, USA
27 – 29 October 2019

ISBN: 978-1-7138-0574-8

AfroBiotech Conference 2019

Atlanta, Georgia, USA
27 – 29 October 2019

AIChE
The Global Home of Chemical Engineers

Printed from e-media with permission by:

Curran Associates, Inc.
57 Morehouse Lane
Red Hook, NY 12571

Some format issues inherent in the e-media version may also appear in this print version.

Printed with permission by Curran Associates, Inc. (2020)

For permission requests, please contact AIChE
at the address below.

AIChE
120 Wall Street, FL 23
New York, NY 10005-4020

Phone: (800) 242-4363
Fax: (203) 775-5177

www.aiche.org

Additional copies of this publication are available from:

Curran Associates, Inc.
57 Morehouse Lane
Red Hook, NY 12571 USA
Phone: 845-758-0400
Fax: 845-758-2633
Email: curran@proceedings.com
Web: www.proceedings.com

TABLE OF CONTENTS

Valorization of Some Solid Environmental Wastes for Treating Liquid Waste: The Closed Switch

Authors:

Unuofin, J., University of Fort Hare

The prospect of enzymatic treatment of pollutants has gradually been receiving growing attention in contemporary times due to the their environmental friendliness and production economic feasibility. Laccase, a multi-copper oxidase has heightened its appeal towards environmental and biotechnological applications due to its broad substrate specificity and its requirement of atmospheric molecular oxygen as a co-substrate and the discharge of water as the byproduct. Hence, this present study was designed to evaluate the biotechnological potentials of laccases produced by some bacteria species from some aquatic milieu of the Eastern Cape Province, South Africa. Optimization of laccase production in a submerged fermentation was by traditional and statistical methods, where the fermentative production of laccase from lignocellulosic agroindustrial residues was evaluated. Laccase secretions produced from statistical optimization exhibited enhanced polyextremotolerant capacities. They were active at a broad range of temperature (0-90°C); with optima at 70°C (Hb9c) and 60°C (Ie1c), pH (3-11); with optima at pH 6 (Ie1c) and pH 8 (Hb9c), respectively, and were equally thermo- and pH-stable. Their activities were either improved or left unabated by high concentrations of cations, detergents, and chloride. In addition, catalytic activities of some secretions increased when they were pre-incubated with 2 " 20% of fluoride, a potent inhibitor. Consequently, a molecular perspective depicted the isolates to have multiple homologous laccase encoding genes. The laccases were remarkable in the decolourization of synthetic dyes. Ultimately, the application of free laccases in denim bleaching, individually or with a blend of a mediator, ABTS, showed that denim colors could be bleached without the need for chemical bleaching agents. The results obtained suggest the bacterial laccases produced from lignocellulosic wastes may serve as potent degraders of phenolic pollutants in water and, may also contribute to the bioeconomy and promote greener techniques for industrial applications.

Cancer Cells Possess Different Isotopic Enrichment

Authors:

Little, R., Stillman College

Although the dynamics of telomeres during the life expectancy of normal cells have been extensively studied, there are still some unresolved issues regarding this research field. For example, the conditions required for telomere shortening leading to malignant transformation are not fully understood. In this work, we mass analyzed DNA of normal and cancer cells for comparing telomere isotopic compositions of white blood cells and cancer cells. We have found that the 1327 Da and 1672 Da characteristic telomere mass to charge cause differential mass distributions of about 1 Da for determining isotopic variations among normal cells relative to cancer cells. These isotopic differences are consistent with a prior theory that replacing primordial isotopes of ^{1}H, ^{12}C, ^{14}N,^{16}O, ^{24}Mg, ^{31}P and/or ^{32}S by nonprimordial, uncommon isotopes of ^{2}D, ^{13}C, ^{15}N,^{17}O, ^{25}Mg and/or ^{33}S leads to altered enzymatic dynamics for modulating DNA and telomere codons towards transforming normal cells to cancer cells. The prior theory and current data are consistent also with a recently observed non-uniform methylation in DNA of cancer cells relative to more uniform methylation in DNA of normal cells.

Design, Synthesis, and Characterization of DNA Origami for the Detection of Biomolecules

Authors:

Stewart, J. M., California Institute of Technology

Rothemund, P. W. K., California Institute of Technology

Analysis of nucleic acid sequences and proteins is a powerful tool in medicine, allowing for the diagnosis of disease and identification of genetic mutations. We propose a platform using DNA origami, a technique for creating nanoscale shapes out of DNA. This DNA origami device is bistable, capable of reconfiguring between two states of 'open' and 'closed' dependent on the presence of an analyte of interest, such as a specific DNA sequence or protein. In solution, this conformational change can be determined by an electrophoretic gel shift. To detect this conformational change on a surface we can observe a structural change by atomic force microcopy. More specifically, we can detect this structural change on an electroactive surface by utilizing the large DNA assemblies to integrate signaling molecules for 30X amplification from a single molecule binding event. These signaling molecules donate electrons at different kinetic rates based on moving the signaling molecule closer or farther away from the surface and can be detected through electrochemical methods. Both binding domains and signaling molecules are integrated into a single DNA origami device eliminating many laborious steps required for a standard immunoassay such as sandwich ELISA, which require several different components to be assemblies and purification. The implications of this work will foster novel innovations for point of care testing for disease.

High-Throughput Cell Programming with Software, Automation and Next Generation Sequencing Technologies

AfroBioTech

Authors:

Habimana, A., Ginkgo Bioworks

High-throughput cell programming with software, automation and next generation sequencing technologies

Aimable Habimana

NGS Test Engineer

Ginkgo Bioworks

The past decade has witnessed a dramatic fall in the cost of reading and writing DNA. These innovations have enabled researchers to design ,quickly test new genes and apply genetic engineering to a variety of new fields. Ginkgo Bioworks, a biological engineering company in Boston, MA, emerged during this period with the mission of using these rapidly advancing technologies to 'make biology easier to engineer'. This presentation will overview the systems Ginkgo has put in place to engineer biology and to answer custom challenges for our customers. We will discuss successful case studies of pathway design in microbes using software and hardware automation, with a particular focus on how Ginkgo™ centralized NGS service exploits using high-throughput library preparation technologies and data analysis pipelines to identify and quantify expression of thousands of desired products. We will also cover how Ginkgo is utilizing foundry capabilities to partner with new customers and explore new markets in agriculture and pharmaceuticals. The talk will close with an appreciation of the people behind the science by touching on the importance of diversity and inclusion in synthetic biology and efforts that Ginkgo is putting in place to improve in this area.

Development of a Lift-Off Intradermal Cancer Vaccine Release System Using Electrostatic Assembly

Authors:

Fletcher, S., Massachusetts Institute of Technology

Melanoma is the fifth most common cancer in men and sixth in women and incidence rates have grown rapidly over the past 30 years.[1] The field of cancer immunotherapy has been recognized as a promising approach to treat this disease, but improvements in response rates are needed to make this treatment more effective for patients.[2] A melanoma vaccine consisting of a nucleic acid adjuvant and protein immunogen is a treatment that can trigger an immune response to cancer. Microneedle technology enables vaccines to be delivered directly to the skin where critical immune cells reside, potentially improving the elicited response.[3, 4] A vaccine can be constructed on the microneedle surface using layer-by-layer assembly, which utilizes the electrostatic attraction between charged materials to create a polymeric film that degrades over time.[4] Our team investigated layer-by-layer assembled silicon chip vaccines and tested them *in vitro* to determine the ideal adjuvant type and influence of film formulation on release kinetics and the stimulated response. Using a tetralayer formulation, sustained release of an RNA adjuvant and a strong immune response was achieved. Future work will involve the translation of the studied films to microneedles and examine their *in vivo* response.

References

1. *Cancer Facts & Figures.*American Cancer Society 2018.
2. Anguille, S.; Smits E. L.; Lion, E.; Tendeloo, V. F.; Berneman, Z. N., Clinical use of dendritic cells for cancer therapy. *The Lancet Oncology***2014**, 15 (7), e257-e267.
3. Glenn, G. M.; Kenney, R. T., In *Mass Vaccination: Solutions in the Skin*. Springer-Verlag Berlin Heidelberg, 2006, pp. 247-268.
4. DeMuth, P. C.; Min, Y.; Huang, B.; Kramer, J. A.; Miller, A. D.; Barouch, D. H.; Hammond, P. T.; Irvine, D. J., Polymer multilayer tattooing for enhanced DNA vaccination. *Nature materials***2013**, 12 (4), 367-376.

MECHANISMS OF CANCER-BONE MICROENVIRONMENTAL INTERACTIONS IN PROSTATE CANCER

Authors:

Odero-Marah, V., Center for Cancer Research and Therapeutic Development, Department of Biological Sciences Clark Atlanta University

Henderson, V., Center for Cancer Research and Therapeutic Development, Department of Biological Sciences Clark Atlanta University

Prostate Cancer (PCa) patients' mortality is mainly attributed to complications caused by metastasis of the tumor cells to organs critical for survival, such as bone. We hypothesized that prostate cancer cell-bone interactions would promote paracrine signaling. A panel of prostate cancer cell lines were co-cultured with Hydroxyapatite (HA; inorganic component of bone) of different densities. Conditioned media (CM) was collected and analyzed for calcium levels and effect on paracrine signaling, cell migration and viability *in vitro* and *in vivo*. Our results showed that calcium levels were elevated in CM from cancer cell-bone co-cultures, compared to media or cancer cells alone, and this could be antagonized by EGTA, a calcium chelator, or knockdown of Snail protein. We also observed increased STAT3 phosphorylation and paracrine cell proliferation and migration in LNCaP cells incubated with CM from various cell lines; this phosphorylation and cell migration could be antagonized by Snail knockdown or various inhibitors including EGTA, STAT3 inhibitor (WP1066) or Cathepsin L inhibitor (Z-FY-CHO). *In vivo*, higher HA bone density increased tumorigenicity and migration of tumor cells to HA implant. Finally, nuclear Snail and Cathepsin L was more highly expressed in cancer compared to normal patient tissue. Our study shows that cancer-bone microenvironment interactions mediated by Snail leads to calcium-STAT3 signaling, which may present an area for therapeutic targeting of metastatic prostate cancer.

Molecular Components and Trafficking of ARRDC1-Mediated Microvesicles(ARMMs)

AfroBioTech

Authors:

Beyene, J., Harvard

The ability to efficiently package and deliver therapeutic molecules to target cells and tissues in diseased individuals remains starkly elusive. Among the many exacerbations that occur with treatment, administered therapeutic cargos often suffer from poor permeability, high toxicity, and often evoke dangerous autoimmune responses. We have recently discovered novel small vesicles known as ARMMs (ARRDC1-Mediated Microvesicles) that are naturally secreted by mammalian cells. We have found that ARMMs exhibit several important features that make them suitable for delivery of biological therapeutics: 1) ARMMs can be readily produced by overexpression of the ARRDC1 protein and are relatively uniform in size (average ~100 nm), 2) specific protein molecules are actively recruited into ARMMs and are protected by the lipid bilayer from degradation, and 3) ARMMs can deliver bioactive cargos into recipient cells. In addition, ARMMs likely elicit little immunogenic response as they are produced by human cells endogenously and can be found in circulating blood.

I have shown that ARMMs can efficiently package and functionally deliver diverse payloads (tumor suppressor protein p53, GFP and p53 RNAs, and the CRISPRCas9/guide RNA complex) into cultured recipient cells, however key questions remain about the endogenous molecular components of ARMMs and the mechanisms of ARMMs uptake into recipient cells. Elucidating these aspects of ARMMs trafficking are paramount to understanding their function and utility as a vehicle for the delivery of biological therapeutic medicines.

Synthesis and Preliminary Biological Evaluation of Highly Substituted Indoles

Authors:

France, S., Georgia Institute of Technology

Highly substituted indoles are among the most prestigious structures in medicinal chemistry. They found in a vast array of biologically active natural products and pharmaceutical compounds. Due to their relevance, methodologies that allow for the regioselective synthesis of indoles are highly desirable. Herein, we would like to discuss our efforts towards the indole core by exploring the chemodivergent reaction of Î±-diazo-Î²-ketoesters and enol ethers. We found that copper (II) catalysts led to the exclusive formation of 1,2-dihydrofuran (DHF) acetals, which served as efficient precursors to the synthesis of indoles. In contrast, Rh (II) catalysts led to the direct formation of regioisomeric indoles. These indoles were then then examined for potential biological activity based on docking models and modified synthetic strategies.

A New Approach to Regenerative Cartilage Tissue Engineering Using Smart-Sensitive Nutraceutical Hydrogels

Authors:

Mendenhall, J., Morehouse College

Injury and diseases that affect articular cartilage present a daunting challenge in orthopedic medicine. During the onset of injury or disease, low oxygen environments decrease healthy cartilage cell growth by limiting extracellular matrix production and creating reactive oxygen species (ROS) such as nitric oxide (NO). Ultimately, this environment reduces the efficacy of traditional biomaterials to cultivate a regenerative scaffold. Our research group has developed a therapeutic injectable hydrogel that provides a regenerative interface for cartilage tissue engineering via a cell-protective mechanism. Using Poly(N-vinylcaprolactam)[PVCL] a smart biomaterial that changes its molecular orientation upon temperature change along with its mechanical properties in conjunction with other natural therapeutic molecules. The hydrogel composite, containing PVCL-hyaluronic acid-NUTR (PVCL-HA-NUTR) affords a robust hydrogel with tunable (lower critical solution temperature) [LCST] parameters near physiological temperature. Our hydrogels have been investigated *in vitro*static cell culture and *in vivo*studies. Using 3D printing, hydrogels were also fabricated to compare the effect of material properties on cell proliferation and metabolism on extra cellular matrix (ECM) proteins. Fetal bovine chondrocytes were harvested and seeded [or 3D bioprinted] into various formulations of PVCL-HA-NUTR under normoxia (21%) and hypoxia (1%) low oxygen conditions. Our results show that chondrocyte cell viability at 1% O_2levels remained higher than that of 20% O_2levels in PVCL-HA for each time point. Animal studies were performed for 30 days after injection of therapeutic hydrogels into the defective knee joint confirm higher amounts of chondrocyte cloning within the murine joint. This work shows promise in providing a cyto-protective biomaterial construct that will begin to regenerate diseased articular cartilage.

Engineering Modular Components of an Osteochondral Implant

Authors:

Matthew, H. W. T., Wayne State University

Miles, K. B., Wayne State University

Vossoughi Shahvari, A., Wayne State University

Alamin, T., Wayne State University

Osteoarthritis remains a poorly manageable outcome of traumatic cartilage injury as cartilage is unable to self-repair. Transplantation of an engineered, osteochondral tissue plug has been proposed as a better way to achieve repair of cartilage defects. Chitosan-based materials that support both chondrogenesis and osteogenesis while promoting rapid tissue integration may provide a viable solution. Bone marrow MSCs were co-encapsulated in glycosaminoglycan-chitosan microcapsules (300-400 μm diameter) with either hydroxyapatite (HAP) microgranules or methacrylated hyaluronan-collagen hydrogels and cultured in either osteogenic or chondrogenic induction media respectively for 4 weeks. Capsule cultures were analyzed to evaluate cell growth and distribution, matrix composition and gene expression. In chondrogenic cultures the effects of hypoxia and medium perfusion were also evaluated, while osteogenic capsules fused into 3D constructs and subjected to compressive testing. In chondrocytic capsules, quantitative assays showed that hyaluronan-collagen inclusion enhanced cell growth and matrix (s-GAG & collagen II) deposition. Growth was further promoted by both perfusion and maintenance in hypoxic (2% oxygen) conditions. PCR results showed that collagen incorporation, perfusion and hypoxia together combined to enhance expression of collagen II, aggrecan and SOX-9 by up to 10-fold. Osteoinduced MSCs deposited significant bone-like mineral in microcapsule interiors, including mineralization of capsule membranes. Compressive testing showed that mineralization increased capsule construct mechanical properties by at least 3 orders of magnitude. HAP-containing capsules had compressive yield strengths of 10 MPa and compressive moduli of 45 GPa. These results demonstrate that MSCs in GAG-chitosan capsules can be cultured to generate in vitro constructs with physical and biological properties approaching those of cartilage and bone and may be suitable for in vivo testing of an osteochondral plug.

Engineering Nano-Delivery for Diagnosis and Restoration of Disrupted Brain Endothelium Post Blast-Induced Traumatic Brain Injury (TBI)

Authors:

Inyang, E., University of Texas, Arlington

Kuriakose, A., University of Texas, Arlington

Chen, B., University of Texas, Arlington

Nguyen, K., University of Texas, Arlington

Cho, M., University of Texas, Arlington

Traumatic brain injury (TBI) associated with explosive blasts is a serious concern among military personnel, potentially causing neurological sequelae. Although disruption of the blood-brain barrier has long been suspected, the coupling mechanisms are yet to be elucidated. Our previous studies have shown that a blast can generate highly pressurized micron-size bubbles, and the microbubbles subsequently collapse and produce a secondary shock pressure (i.e., microjet). Using tissue engineering approaches, we recently demonstrated that the microjet can cause disruption of brain endothelium (BE) tight junctions. High expression of E-Selectins is typically indicative of inflammation and injury to endothelial cells (ECs). Upregulation of this protein following an injury can be exploited for diagnosis and potential therapy through targeted delivery. We hypothesized that BE disrupted by microject could be targeted, diagnosed, and treated for repair and restoration. To test this hypothesis, we engineered nanoparticles (NPs) to encapsulate therapeutic reagents and decorated the NPs with peptides to specifically bind to the damaged BCs.

Poly (lactic-co-glycolic acid; PLGA) nanoparticles (NPs) were successfully fabricated and characterized. The PLGA NPs were functionalized with the ligands (CD162) to target E-Selectins expressed on the damaged ECs. Damaged cells exhibited a high affinity to conjugated NPs compared to unconjugated control NPs. Such cellular uptakes are quantified by measuring the accumulation of intracellular NPs. The cellular repair was quantified by measuring the rate of proliferation of brain ECs to restore the injury site. Targeted delivery of non-biologics as potential therapeutic reagents is developed and tested to repair and restore the BE.

Exploiting the tunable properties of PLGA, a targeted drug delivery strategy was developed for targeted treatment of injured brain ECs. Because the current approach combines the specificity of ligand/receptor interaction with therapeutic reagents, it provides a potential theragnostic approach for treatment of TBI.

Increased Hypercoagulation Markers in a Humanized Sickle Cell Mouse Model Caused By Activated Cathepsin-Mediated Fibrinogenolysis

Authors:

Douglas, S., Georgia Institute of Technology & Emory University

Ahmed, O., Georgia Institute of Technology & Emory University

Botchwey, N., Georgia Institute of Technology & Emory University

Song, H., Georgia Institute of Technology & Emory University

Platt, M., Georgia Institute of Technology & Emory University

Sickle cell disease (SCD) affects approximately 100,000 people in the United States. Globally 300,000 babies are born with SCD. Coagulation is chronically activated in SCD, and etiology is not understood, but can contribute to sickle pain crisis. We need to understand mechanisms underlying chronic coagulation in SCD to develop anticoagulant therapies for the SCD population. During coagulation, thrombin cleaves fibrinopeptide A (FpA) and B from fibrinogen. People with SCD have increased FpA in plasma, possibly contributing to chronic coagulation activation. Computational docking identified cathepsin K (catK) has cleavage sites on fibrinogen near a Gly-Pro-Arg sequence, releasing FpA. Inflammation of SCD causes increased catK production by vascular cells, and we hypothesize that increased plasma levels of FpA in SCD is due to catK activity. We assessed amounts of active cathepsins and FpA in plasma of humanized sickle cell mice: non-sickle (AA), sickle cell trait (AS), and sickle cell anemia (SS) genotypes. To determine if cysteine cathepsins were causal agents of increased FpA, mice were injected with E64, cathepsin inhibitor, from ages 1 to 3 months, hypothesizing that E64 will reduce catK activity, and generate less FpA. From zymography, untreated 1 month SS mice had twice as much active cathepsins compared to 1 month AA and AS mice. Plasma was elevated in 3 month SS mice compared to 1 month SS mice. SS male mice had more active cathepsins than SS females. Preliminary FpA ELISAs show untreated SS mice may contain higher amounts of circulating FpA compared to AA and AS mice, but E64 treatment reduced amounts of active cathepsins and FpA. These data support that SCD mediates increased levels of active catK in the plasma which cleaves FpA from fibrinogen. We are among the first to show this in an animal model, which helps us understand chronic coagulation activation in SCD.

You Are What You Eat: The Impact of Obesity on B-ALL Pathogenesis and Therapeutic Responses

Authors:

Henry, C., Emory University

Relapsed B-cell acute lymphoblastic leukemia (B-ALL) is the second leading cause of childhood cancer with survival outcomes declining by 30% in obese relative to lean pediatric patients receiving chemotherapy treatment. Poor survival outcomes in obese pediatric patients with B-ALL has largely been attributed to altered chemotherapy pharmacokinetics; whereas, the direct impact of the obese microenvironment on B-ALL cells and T-cell immunity has been largely unexplored. We have found that adipocytes directly induce chemoresistance in murine and human B-ALL cells. Furthermore, we have observed compromised T-cell function in murine models of obesity and in obese patients with B-ALL. Mechanistically, we found that adipocytes induce Galectin-9 (GAL-9) surface expression on B-ALL cells which promotes chemoresistance and attenuates T-cell mediated immunity. We have discovered that antibody-mediated blockade of GAL-9 promotes extensive cell death in adipocyte-conditioned B-ALL cells *in vitro* and significantly prolongs the survival of obese mice transplanted with B-ALL cells relative to those receiving chemotherapy treatment. Results from these studies highlight a novel approach to overcome obesity-induced chemoresistance and serve as the foundation for engineering chimeric antigen receptor (CAR) T-cells expressing anti-GAL-9 antibodies to improve survival outcomes in obese patients with B-ALL.

Engineered Autonomous Control of Metabolic Pathways

AfroBioTech

Authors:

Prather, K. L. J., Massachusetts Institute of Technology

Microbial systems offer the opportunity to produce a wide variety of chemical compounds in a sustainable fashion. Economical production, however, requires processes that operate with high titer, productivity, and yield. One challenge towards maximizing yields is the need to use substrate for biomass, resulting in a competing pathway that cannot merely be eliminated. Productivities may also be significantly influenced by the timing of expression of genes in the production pathway. Dynamic metabolic engineering has emerged as a means to address these and other impediments in strain performance. Ideally, the triggers for dynamic control would be autonomous, that is, independent of any external intervention by the operator. We have developed such autonomous devices by utilizing both pathway-independent quorum-sensing circuits and pathway-dependent metabolite biosensors to control metabolic flux. In this talk, I will describe our approach for development of these Metabolite Valves and results to date from their implementation.

Multiplexed Microanalytical Biosensor Systems

Authors:

Guiseppi-Elie, A., ABTECH Scientific, Inc.

A wide range of nano-enabled materials chemistries results in the synthesis of hybrid, multifunctional, bioactive, and electroconductive hydrogels that enable control of the ABIO-BIO interface of indwelling multiplexed analytical microsystems. This paper surveys ongoing research at the Center for Bioelectronics, Biosensors and Biochips (C3BÂ®) that is aimed at: i) A penta-analyte biosensor, the PSM BioChip, for the monitoring of key biomarkers of physiological stress during trauma-associated hemorrhage. This intramuscularly indwelling microsystem measures lactate, glucose, potassium, oxygen tension and pH (acidosis). The system uses decision tree, artificial neural networks and expert data to fuse sourced analytical data into a single real-time, reportable, actionable score useful in mass triage, trauma resuscitation management and ICU monitoring. ii) A dual-function 24-well Electrical Cell Stimulation And Recording Apparatus (ECSARA) for the concomitant electrification of human inducible pluripotent stem cells (hiPSCs) to guide differentiation while monitoring and modeling transmembrane epithelial / endothelial cell resistance (TEER) and capacitance of the cells as a tool to evaluate cellular behavior. This system applies electric fields of relevant physiological values to cells grown in culture on microporous inserts and does so for long periods with scheduled intermittent measures of electrical impedance spectra. iii) A dual-responsive biosensor system that uses biologically responsive organic electronic circuit elements for the fused monitoring of glucose and lactate levels reported wirelessly via the output of an oscillator circuit. iv) A biosensor-enabled, dual-sensing microfluidic system that recapitulates vascular physiology for the investigation of flow fields, agonists and antagonists on human inducible pluripotent stem cells (hiPSCs) derived endothelial cells (ECs) from patients with Hutchinson-Gilford progeria syndrome (HGPS). This microfluidic vasculature-on-a-chip system measures nitric oxide (NO) and reactive oxygen species (ROS) as key indicators of vascular response to stress.

Modeling and Predicting the Behaviors of Cytoskeletal Networks

Authors:

Belmonte, J. M., North Carolina State University

Cytoskeletal networks drive many cellular and morphological processes such as spindle organization and apical constriction. Although cytoskeletal components and their properties are known, we still lack an intuitive understanding of how cytoskeletal networks behave. Here, I describe an analytical theory that predicts whether a network will contract, expand, or conserve its dimensions. This analytical theory correctly predicts the behavior of simulated networks, consisting of filaments with varying combinations of connectors and motors. We find that contraction is the dominant behavior of networks with semi-flexible filaments, such as F-actin. By contrast, networks of rigid filaments, such as microtubules, can be either contractile or extensile, and the theory reveals unexpected modes for network expansions that do not depend on *de novo* assembly of filaments. Our results suggest that pulsatility is an intrinsic behavior of contractile networks if the filaments are not stable but turn over. The theory offers a unifying framework to think about mechanisms of contractions or expansion and allows for a qualitative prediction of their time-evolution. It provides the foundation for studying a broad range of processes involving cytoskeletal networks and a basis for designing synthetic networks.

Towards Real-Time in Vitro cell Culture Monitoring: Detecting Changes in LN-18 Glial Cell Morphology Using a Supervised Machine Learning Approach

Authors:

Gilmore, J., Clemson University

Mbiki, S., Clemson University

McClendon, J., Clemson University

Samec, T., Clemson University

Alexander-Bryant, A., Clemson University

In cell-based research, the process of visually monitoring cell cultures generates large image datasets that need to be evaluated for quantifiable information in order to track effectiveness of treatments *in vitro*. However, the traditional approach to annotating image data is tedious and error prone. This project sought to create an image classification framework that employs machine learning to detect different LN-18 glial morphology. Cell images captured from the EVOL-L microscope system underwent preprocessing to enhance contrast and reduce noise. A training set was created using labeled images and extracted shape and texture features. A non-linear SVM with a Gaussian Radial Basis kernel was trained using the training set. The resulting classifier was tested using a test set containing new images. The testing period showed that the classifier had a classification accuracy of 76%. The trained classifier was then applied to a database containing 636 cell images.

Varying PLGA and Peg Oligomer Extension Reveals Favorable Protein/Polymer Interactions Critical to Drug Loading of PLGA-Peg Nanoparticles

Authors:

Nyambura, C., University of Washington

Pfaendtner, J., University of Washington

Nance, E., University of Washington

Protein therapeutics have the potential to treat a wide range of life-threatening diseases in affected patient populations. When compared to small molecule drugs, this class of macromolecules can provide high specificity and complexity in their function and replace critical proteins in individuals missing necessary genes or have mutated genes containing key structure and function information. Despite these advantages, protein drugs can cause adverse effects, depending on the route of administration. Such effects include protease recognition and cleavage once in the bloodstream, development of cross-reactive neutralizing antibodies and an increased risk of injection-site or viral infections. Polymer nanoparticles can better ensure translation of these therapeutics to the clinic by protecting them against degradation and immune recognition, controlling their release, assist in crossing biological barriers and by targeting specific sites of action. Nanoparticles (NPs) comprised of polylactic-co-glycolic acid polyethylene glycol (PLGA-PEG) are non-toxic, non-immunogenic. Using this diblock copolymer, desirable nanoparticle characteristics, such as neutral surface charge and uniformity in size and dispersity, can be achieved but only after extensive manipulation of formulation parameters. Moreover, experimental investigations of molecular-level behavior during nanoparticle synthesis are few, due to detection limits and/or lack of instrument sensitivity. Therefore, this work uses restrained, atomistic molecular dynamics (MD) to examine the role of polymer extension in the protein encapsulation process within PLGA-PEG NPs. We simulated three PLGA (N = 20 monomer length) and three PEG (N = 20 monomer length) polymer oligomers in the presence of a model protein, bovine serum albumin, and an explicit water solvent medium. A harmonic potential was used to restrain the oligomers' radius of gyration (R_g) at various values reflecting different levels of extension arising from varying solvent quality. Results indicate that protein-polymer contacts are promoted at intermediate values of polymer R_g for both PLGA and PEG.

Controlling Metastasis and Immune Signaling in Breast Cancer Cells with Epigenetic Engineering

Authors:

Haynes, K., Emory University

Baskin, N., Emory University

The role of chromatin-driven epigenetics in drug resistant cancers is becoming increasingly clear with the growing body of genome-wide expression data. Epigenetic therapy, a class of drugs that block enzymes that regulate protein-DNA structures (chromatin) in the nucleus, is gaining traction as a way to reactivate silenced anti-cancer genes in cancer cells. Triple negative breast cancer (TNBC) disproportionately affects African-American women and shows high rates of relapse after chemotherapy. TNBC responds moderately to leading epigenetic drugs *in vitro, in vivo*, and in clinical trials. Improvement of epigenetic therapy for TNBC is impeded by a lack of chromatin-targeting pharmacological mechanisms. Inhibition of hyperactive EZH2, which leads to depletion of histone H3 lysine 27 (H3K27me3), passively affects gene regulation, affects cellular processes outside of chromatin, and is therefore not ideal for targeted epigenetic control. Our goal is to engineer histone-reader-effector proteins, which are central components of the natural chromatin machinery. We set out to determine the efficacy of a H3K27me3-binding transcriptional activator, Polycomb-based Transcription factor (PcTF), in TNBC cells compared to small molecule inhibitors of EZH2. In order to determine the pharmacological potential of PcTF and inhibitors, we have designed XTT experiments to measure IC50 values for several breast cancer cell lines. So far, we have observed varying sensitivities of TNBC cells to a panel of EZH2 inhibitors. We have developed a cell-penetrating PcTF to support dose-response experiments to generate results (IC50) that can be compared with inhibitor treatments. We expect that PcTF will show anticancer effects since we previously demonstrated that transgenic over-expression of PcTF activates genes that reverse the epithelial-mesenchymal transition (EMT) and stimulate the T-cell signaling pathway. In summary, we have established a fusion protein-based strategy to expand epigenetic therapy beyond small molecule inhibitors. Efforts to extend our viability (XTT) analysis to apoptosis, invasion, and metastasis are underway.

Enabling tools for metabolic engineering in anaerobic fungi

AfroBioTech

Authors:

Solomon, K. V., Purdue University

Early-diverging anaerobic fungi (phylum Neocallimastigomycota) are attractive emerging platforms for metabolic engineering due to their robust degradation of untreated plant feedstocks and novel biosynthetic capabilities. More importantly, anaerobic fungi are naturally competent during portions of their lifecycle and are genetically tractable. To effectively leverage these abilities, however, an engineering toolbox is required. My lab is pursuing a number of plasmid-, genomic-, and epigenomic-strategies for tool development to domesticate anaerobic fungi for biotech. We have identified several gene regulatory elements from transcriptomic and genomic resources to drive gene expression for the first time in either the nucleus or cytoplasm of anaerobic fungi. Validation studies with flavin-based fluorescent reporters and selection markers confirm sequence dependent increases in gene expressing in a controllable manner. These tools enable library-based screening approaches to construct the first stable anaerobic fungal plasmid and the introduction of CRISPR-based endonucleases for genomic integration. We have also pioneered epigenetic regulation of anaerobic fungi and identified several chemical inhibitors that can perturb the epigenome of anaerobic fungi with concurrent phenotypic changes in biomass-degrading enzyme secretion. Efforts are currently underway to capitalize on systems biology methods to rapidly elucidate and expand parts available for engineering applications. In parallel, we have also examined the feasibility of anaerobic fungal process development with partner microbes such as *K. marxianus*. Anaerobic fungi readily release a suite of fermentable sugars and organic acids that support the growth of engineered *K. marxianus* to make value-added esters and fragrance compounds from crude biomass. These examples showcase a few of the metabolic engineering opportunities currently available with anaerobic fungi. More importantly, our growing toolbox of genetic and epigenetic strategies can be readily generalized to other non-model systems, and will soon enable direct metabolic engineering in anaerobic fungi.

Mating Disruption: The New Standard for Insect Control

Authors:

Glenn, W., Dr.

First, this talk will survey the benefits of mating disruption as a means of controlling insect populations near crops of interest. Traditional methods to control insects heavily rely upon seven broad classes of insecticides, some of which pose toxicity risks to farmers and consumers. Moreover, some traditional methods have been over-utilized, increasing the likelihood of resistance. By contrast, lowering insect populations through mating disruption is not toxic and is not subject to resistance. Further, we note that adopting mating disruption technology can potentially extend the lifespan of effective low-toxicity insecticides when incorporated into a diversified integrated pest management portfolio.

Next, this talk will explore the utility of mating disruption technology in existing markets, which include nuts, pome fruits and stone fruits. Further, this talk will also explore how Provivi is creating new markets for mating disruption in staple row crops including rice and corn, which supply nearly half the calories consumed across the globe each day.

Finally, this talk will highlight various strategies to synthesize the pheromones that enable the mating disruption technology.

We Stress Bacteria out: Dynamic Responses of Bacteria to Fluid, Nanoparticle, and Antibiotic Stress.

AfroBioTech

Authors:

Jones, A. A., Northeastern University

Our engineered environments concentrate and put bacteria under a lot of stress. However, we know for humans that there is good and bad stress. We explore how three different stressors and attempt to classify those stress responses to help both disease treatment and bioprocessing. We show that increased shear stress increases the metabolism of electroactive active bacteria independently of mass stress in a pure culture biofilm. We use a rotating disc electrode to decouple mass transport from shear stress. We find increased shear stress reduces biofilm development time while increasing its metabolic rate. Furthermore, biofilm health is negatively affected by higher metabolic rates over long-term growth due to the biofilmâ€™s memory of the fluid flow conditions during the initial biofilm development phases.

We also show that bacteria growth curves can be used to predict resistance or tolerance development when bacteria are exposed to nanoparticles and antibiotics. We use machine learning approaches to rapidly classify survival behavior across nanoparticle types and concentrations. We compare 1000 experimental interactions between *Escherichia coli*, *Staphylococcus aureus*, *Staphylococcus epidermidis*,*Pseudomonas aeruginosa*,*Helicobacter pylori*and concentrations of liposomal drug delivery systems, amphiphilic peptide, and silver and selenium nanoparticles. Clustering reveals bacteria-nanoparticle clusters separate resistance and tolerance development from stable growth. Furthermore, bacteria-generated nanoparticles do not induce growth dynamics associated with resistance.

Dual oxidase1/Lactoperoxidase Antimicrobial System Deplays Anti-Viral Effects in Cell-Free and Cell-Based in Vitro and Ameliorate Survival In Vivo during Influenza

Authors:

Sarr, D., University of Georgia

Gingerich, A., University of Georgia

Colon, M., University of Georgia

Nagy, T., University of Georgia

Tripp, R., University of Georgia

Rada, B., University of Georgia

Chandler, J., Emory University

Influenza virus infections cause severe morbidity and mortality, especially in high-risk patient populations. Broncho-epithelial cells orchestrate an oxidative antimicrobial system present in the airway liquid consisting of lactoperoxidase, the thiocyanate ion, and hydrogen peroxide. Dual Oxidase 1, an enzyme highly expressed in these cells, is the source of hydrogen peroxide, which in the presence of lactoperoxidase, catalyzes thiocyanate to hypo-thiocyanate ions with anti-viral properties.

Our objective was to demonstrate the effectiveness of this system using a cell-free and cell-based *in vitro* as well as *in vivo*. Our *in vitro* studies used a combination of both components of the system and assessed whether the full system has anti-viral effects. In a cell-based system, components were combined with the exception of hydrogen peroxide, which was expected to be generated by Dual Oxidase 1. Using a mouse adapted influenza strain PR8, we tested whether Dual oxidase 1 has a protective role *in vivo* using *Dual oxidase 1* null mutants and wild type littermates mice. Our data demonstrate that the system has anti-viral effects in both cell-free and cell-based *in vitro*. While *in vitro* studies suggest an anti-viral properties of hypo thiocyanate, the *in vivo* effect of the system has never been addressed to our knowledge. *Dual oxidase 1*-deficient mice have significantly increased morbidity and mortality following influenza infection compared to controls. The characterization of infiltrated leukocyte subsets following infection revealed impaired natural killer cell recruitment in the lung of *Dual oxidase 1*-deficient animals.

In summary, we provide evidence of anti-viral effects of hypo-thiocyanate *in vitro* in cell-free and cell-based systems and demonstrate the *in vivo* antiviral role of Dual oxidase 1. All together, improving this system *in vivo* has the potential to provide novel treatment or prophylactic strategies against influenza and could be extended to other respiratory infections.

Multifunctional Nanoculture Systems for Studying Microbial Dynamics

Authors:

Niepa, T. H. R., University of Pittsburgh

Microbes self-organize at various interfaces while transitioning to a sessile form within a protective biofilm matrix. While the biological implications of biofilms for the environment, health, and industry are widely appreciated, the earlier developmental stage of microbes as microcolonies has received scant attention. Here, we present a new approach to investigate microbial dynamics in spatially confined microsystems. Using microfluidics-enabled fabrication, a nanoliter-scale sessile culture system (*the nanoculture*) is designed to grow synthetic microbial communities. Each nanoculture begins as a several nanoliter droplets of suspended cells, encapsulated by a polydimethylsiloxane (PDMS) membrane. The physicochemical properties of the encapsulation materials allow the diffusion of functional molecules to interrogate cell physiology under chemical insults, allowing microbial interactions to be probed within or across the confining vessel. Alternatively, multiple species of microbes can be co-cultured within a nanoculture. Because chemical communication across the membrane occurs, this system can be used to decouple the effects of the physical and chemical interactions between cells, and investigate microbial pathophysiology or bacterial-fungal (inter-kingdom) dynamics that play a central role in early childhood dental caries and many infections. Such a culture system now provides unique opportunities to assess their therapeutic use for personalized medicine, or for high throughput screening of unculturable microbial species relevant to biotechnology and drug discovery.

Quantifying the Host-Pathogen Interface - A Multiscale Perspective on Infection, Immunomodulation, and Disease

Authors:

May, E., University of Houston

Infection by intracellular pathogens such as *Mycobacterium tuberculosis* (Mtb), host immune response, and the occurrence of disease can be viewed as an emergent property resulting from the interaction of multiple molecular and cellular systems. While the tendency is to study microbial pathogens and their host targets individually or from a host or pathogen centric skew, the nature and dynamics of the interaction between the systems are key determinants in infection, immunity, and disease outcome. With the goal of integrating host and pathogen interactions, we develop multiscale experimental and theoretical models to explore pathogen adaptation to environmental stress, and chronic conditions plus comorbidities such as vitamin D deficiency or alcohol use that differentially modulate immune response to infection. Exploiting confocal imaging and 3D tissue culture models, our studies of infection are used to investigate both chemical and structural impacts of host-pathogen dynamics on the clearance of infection, disease onset and persistence. Results demonstrate the importance of host comorbidities, physiology, microenvironment, and biochemical dynamics in the modulation of infection outcome.

GRANT SUPPORT: Research supported in part by DTRA/HJF # FA8650-10-2-6062/subaward 2381, NIH/NHLBI award K25HL075105, and NSF MCB-1445470.

The PTS System and Vibrio Cholerae Biofilms: Analysis, Control, and Applications

Authors:

Ymele-leki, P., Howard University

Phosphoenolpyruvate-carbohydrate phosphotransferase system (PTS) is a highly conserved, multistep chemical process which uses phosphate transfer to regulate the intake and use of sugars and other carbohydrates by bacteria. In addition to controlling sugar uptake, the PTS regulates several bacterial cellular functions such as chemotaxis, glycogen metabolism, catabolite repression and biofilm formation. Regulation of biofilm formation by the PTS is complex, involving many different regulatory pathways that incorporate distinct PTS components. Our studies aim to investigate the synergistic role of enzymes II in the uptake of carbohydrates and the regulation of *Vibrio* biofilms. Additionally, our studies aim to identify and characterize small-molecule modulators of the bacterial PTS functions. Given the connection between PTS activity, biofilm formation, and virulence in *Vibrio cholerae*, we hypothesize that a thorough characterization of the bacterial PTS as well as the identification of novel biological targets will reveal new tools and approaches to regulate biofilm formation by interfering with microbial metabolic functions.

A Novel Fibrous Collagen Scaffold for Patterning Vascular Development and Regenerating the Myocardium

AfroBioTech

Authors:

Minor, A., Brown University

Coulombe, K. L. K., Brown University

Promoting angiogenesis in engineered cardiac tissue is a significant challenge that limits our ability to remuscularize injured myocardium. Biochemical cues from growth factors and cytokines sequestered in the extracellular matrix (ECM) stimulate wound healing and angiogenesis but the distribution of these signaling molecules in biomaterials does not yield spatial control over vessel formation. We aim to develop a heparin-conjugated, patterned, collagen microfiber scaffold to sequester and prolong the release of growth factors in vivo and direct vascular development in engineered human cardiac tissue. Ongoing work uses 1-ethyl-3-(3-dimethylaminopropyl)-carbodiimide (EDC) chemistry to covalently link heparin (Hep-COOH) to type I collagen in solution. EDC and N-hydroxysuccinimide (NHS) were added to a 2% (w/v) solution of heparin in 0.05M MES buffer (pH 5.60) at a molar ratio of EDC : NHS : Hep-COOH of 0.4 : 0.24 : 1.0. Collagen I (13mg/mL) was equilibrated then added to the EDC/NHS activated heparin solution. Fourier-transform infrared spectroscopy (FTIR) revealed S=O asymmetric stretching at 1230cm^{-1} and S-O-C stretching at 990cm^{-1} and ~800cm^{-1} on heparin-modified collagen, suggesting that heparin was conjugated to collagen via carbodiimide chemistry. Heparin-modified collagen will be used to synthesize wet-spun collagen microfiber scaffolds and bind growth factors, providing both spatial and temporal control over the position and release kinetics of growth factors in vivo. Scaffolds will then be embedded in bulk collagen hydrogel with human induced pluripotent stem cell (hiPSC)-derived cardiomyocytes to create a composite material conducive to tissue remodeling and cell infiltration. We expect these patterned microfiber scaffolds to direct host endothelial cell migration and vascular development in a rat ischemia/reperfusion (I/R) model. This approach will allow us to harness native capabilities of host vascularization to increase vessel penetration and architecture of implants in vivo.

Diffusion-Based Modeling and Drug Delivery of Gasotransmitters from Electrospun Scaffolds

Authors:

Washington, K. S., Florida Institute of Technology

Rottmann, S., Florida Institute of Technology

Jourdain, R., Florida Institute of Technology

Abeyrathna, N., Florida Institute of Technology

Liao, Y., Florida Institute of Technology

Bashur, C. A., Florida Institute of Technology

Delivery of gasotransmitters including carbon monoxide (CO) from scaffolds provides local-controlled delivery of cell-signaling molecules to a diseased area that may improve endothelialization in the tissue engineered vascular graft. However, it is important to understanding the proper dose of CO into the cells to determine the endothelial cell response because endogenous levels of CO are anti-inflammatory and required for vascular function, high levels are toxic, and levels in between demonstrate seemingly conflicting results in various tissues [1]. Our previous studies included incorporating CO releasing molecules (CORMs) within electrospun scaffolds and determining the impacts of CO on endothelial cells, where we found the need to better understand CO dose available to the cells. This study aimed to develop a two-part theoretical model designed to simulate diffusion of CO and other traditional drugs from fibrous scaffolds, and determine the temporal concentration of drug available to cells. Our modeling system included: 1) one model that generates the simulated fibrous scaffold structure, and 2) a model that simulates the transport from the fibrous representative volume. For experimental validation, poly(Îµ-caprolactone) (PCL) was electrospin with 2%w/w (FITC)-incorporated and CORM-loaded scaffolds. Samples were prepared and the fluorescence was measured for up to 16 hours to determine the drug release. The model can be used to predict drug availability to cells for a variety of scaffolds and drug molecules. For CO release, this model is necessary because of the limitations with real-time experimental analysis. Results suggest that only a fraction of the initial concentration of gasotransmitters released from fibers that enters the interstitial fluid *in vivo*, or culture media *in vitro*, will be available to cells. We concluded that fiber diameter and fiber density are important parameters not just for traditional tissue engineering, but also for drug delivery.

[1] Washington et. al., *Frontiers in Pharmacology*, 2017.

Manipulating Immune Cells for Tissue (Re)Generation

Authors:

Moore, E., University of Florida

What role can immune cells play in tissue regeneration? My work seeks to answer this question by developing in vitro models of immune cells in blood vessel development as well as studying these immune cells in vivo within injury models. The first part of my talk will focus on the roles of macrophages, cells of the innate immune system, in blood vessel development for tissue generation. We found that M2 macrophages can promote blood vessel development in an in vitro poly(ethyelene glycol) hydrogel. The second part of my talk will introduce the roles of B cells in injury. We found that B cell phenotypes vary based on the biomaterial implant. Pro-regenerative biomaterials are capable of B cell phenotype stimulation.

Regenerative Engineering of the Nail: The Development of a Novel Nano-Based Matrix

Authors:

Montgomery, A., UConn Health

McClinton, M.D., A., UConn Health

Walker, J., UConn Health

Barajaa, M., UConn Health

Washington, Ph.D., K. S., UConn Health

Nair, M.Phil., Ph.D., L., UConn Health

Laurencin, M.D., Ph.D., C. T., UConn Health

Acute traumatic nail injury treatment repair procedures are commonly conducted in emergency departments and primary care offices. Current repair methods use nail splints which are inserted within the nail root to prevent the fusion of the proximal nail fold and the matrix tissue. Splints also provide a protective barrier overlying the nail bed soft tissue during recovery periods, but uncertain prognoses (i.e., aesthetic and functional disadvantages) reveal a need for improvement in nail repair techniques [1]. In this study, we used a coaxial electrospinning method to produce hybrid poly(lactide-*co*-glycolide) (PLGA) (85:15) and gelatin nanofibers to mimic the fibrous nature of the nail's extracellular matrix. Optimal conditions to yield Hybrid$_1$ (PLGA shell, gelatin core) and Hybrid$_2$ (gelatin shell, PLGA core) fibrous matrices showed average fiber diameters of 540 Â± 118 nm and 2215 Â± 1135 nm, respectively. Hybrid$_2$ nanofibrous matrices were crosslinked at 5, 15, and 25% glutaraldehyde (GA) vapor concentrations and results revealed a decrease in average fiber diameter as GA concentration increases. Nail stem cells (NSCs) were positive for stem cell markers CD90 and CD29 and NSC marker Lgr6. K15, K14, and *Anpep* (CD13) expression was found in proximal matrix tissue, however, consistent CD13 expression was observed when NSCs were cultured with 0 mM, 0.15 mM, and 1.87 mM of $CaCl_2$ conditioned media. *In vitro* bioactivity observation of hybrid scaffolds loaded with nail stem cells (NSCs) isolated from rat proximal matrices were viable and showed preferred attachment to Hybrid$_2$ matrices at 24 hours. A regenerative engineered nail matrix may aid to improve cosmetic appearance and function of injured nail organ and the development of suitable animal models for nail regeneration will allow for study of *in vivo* performance. Future work will evaluate subcritical nail injury models along with regenerative engineering strategies to regenerate the nail matrix *in vivo*.

AUTHOR INDEX

AIChE
120 Wall Street, FL 23
New York, NY 10005-4020

ISBN 978-1-7138-0574-8

9 781713 805748

CURRAN ASSOCIATES INC.
proceedings.com